To Irmgard –
a "Special Person"

The Seven Planets

from Anne.

Adam Bittleston

The Seven Planets

Floris Books

First published in volume form in 1985 by Floris Books, Edinburgh. Revised from articles first published in the *Christian Community Journal,* 1951, 1967, 1970.

British Library Cataloguing in Publication Data

Bittleston, Adam
The seven planets.
1. Planets
I. Title
523.4   QB601

ISBN 0–86315–021–7

Printed in Great Britain
at the University Press, Oxford

Contents

Introduction

The articles on the planets collected here originally appeared in a periodical, *The Christian Community,* with a rather specialized readership. Now that they are being offered to a wider circle of readers a few matters may need to be clarified at the start.

It is widely assumed at the present time, particularly in the West, that the only realistic approach to events in the universe, beyond the atmosphere of the Earth, is a purely quantitative one, concerned with large-scale or small-scale *movements* of many kinds. It is not regarded as impossible that we should meet with manifestations of intelligence from beyond the Earth — but it is believed that humanity has not done so yet, and cannot be sure of ever doing so in the future.

Everything in this book is written with a very different conviction — that the heavens can be appreciated in their *qualities* and indeed that there have always been human beings who mastered ways leading to results just as reliable and realistic as quantitative methods. And that bridges can be found between all that is expressed numerically in the observation of the world, and all that we can learn from those kinds of attentiveness through which the soul of the world speaks to us of its mysteries, just as human beings can tell us about the joys, sorrows, and endeavours of their lives.

In the ancient world there were many names for those best able to develop these. Perhaps the most familiar is the name initiate, taken from the Latin, which originally meant something like 'master knower of the beginnings'. The Greek text of the Gospel of St Matthew uses for the 'wise men of the East' the Persian word *Magi.*

In the present century many who have studied his work patiently and open-mindedly have come to regard Rudolf Steiner (1861–1925) as a deeply reliable initiate, accustomed to all the patient exactness of scientific research, and thoroughly conscientious in his use of spiritual faculties of knowledge. His work has been little studied in academic circles, largely because it is so wide-ranging. But it has been used in many countries in practical ways since his death, and increasingly in the last ten years, particularly in the field of education. He provided many keys to the ancient wisdom writings, including the Old and New Testaments, and the Hindu scriptures. His teaching gives a basis for a renewal of Christianity consistent with *all* the major religious traditions which have been passed on from the ancient world.

Recognizing this, a band of men and women founded in 1922, under Rudolf Steiner's guidance, a religious body, The Christian Community. Its task was to celebrate, for the main turning points of life, the original Christian sacraments in a new form. Among them, like a Sun, a form of the Eucharist which takes into account a sensitivity for the soul of nature through which man's present alienation from the earth can be overcome. This is known as the 'Act of Consecration of Man'.

What has been included in this little book has been inspired by the anthroposophical teaching of Rudolf Steiner, and by life as a priest in The Christian Community. But it is hoped that it can serve as an introduction to some mysteries of the Sun and Moon and of those planets which can be seen without the help of a telescope, for readers who have no detailed knowledge either of Anthroposophy or of The Christian Community, but who wish to renew in themselves a love for the heavens which is hidden deep in all our hearts.

1. The Sun

In Edith Sitwell's poem *The Bee-Keeper,* the bees praise earth and water and fire and air. And then they turn to Sun and thunder:

This Sun is the honey of all Beings, and all Beings
Are the honey of this Sun . . . O bright immortal
    Lover
That is the sun and is our Being's sight —
O bright immortal Lover Who is All!

This Thunder is the honey of all Beings, and all
    Beings
Are the honey of this Thunder . . . O the bright
    immortal Lover,
That is in thunder and all voices — the beasts'
    roar —
Thunder of rising saps — the voice of Man!
O bright immortal Lover Who is All!

This was the song that came from the small span
Of thin gold bodies shaped by the holy Dark. . .

This song is founded by Dr Sitwell on one of the Upanishads. It is included in all sincerity in a twentieth-century poem. But can it be *read* with honesty, unless we are willing to face as a great and tragic riddle the condition of human knowledge about the Sun?

The old pre-Christian mysteries culminated in knowledge about the Sun; and this knowledge was concerned with beings who had a great deal to do with the human heart. Though the mysteries declined, and the Christianity of the Church took their place and did not often speak of the Spirit of the Sun, the artists would represent Sun and Moon above the Cross as immediately concerned in the event of Golgotha. Right up to the beginning of the nineteenth century, there was a strong feeling that the worlds of the Sun and stars were beyond the grasp of ordinary earthly thought. The French positivist philosopher Comte could say in 1823 that there was one secret which would be for ever hidden from mankind — the chemical composition of the heavenly bodies. There still seemed plenty of room for poets to speak about the Sun — as Goethe did magnificently in both parts of Faust, or Shelley in *Prometheus Unbound:*

> Ye Kings of suns and stars, Daemons and Gods,
>   Aetherial Dominations, who possess
> Elysian, windless, fortunate abodes
>   Beyond Heaven's constellated wilderness. . .

Yet at the very time that Goethe and Shelley wrote, or Comte assigned this limit to the rational powers he valued otherwise so highly, Fraunhofer in Munich was studying the artificial rainbow of the laboratory, the spectrum, and observing particularly not its colours in their qualities, but its *shadows* — the lines which have taken his name.

Fraunhofer had been rescued, as a fourteen-year-old orphan, from a tenement building which collapsed, killing all its other occupants. Assisted by the Elector of Bavaria, who was passing in his carriage when the disaster happened, Fraunhofer became the supreme lens-maker of his time, supplying the observatories of Europe. He died in 1826, at 39, having established the starting-point of modern astro-physics — that the lines on the spectra of the Sun and individual stars are

1. THE SUN

different, and thus appear to be independent of the Earth's atmosphere.

No considerable progress was made until thirty-three years after his death — when in 1859 Kirchhoff at Heidelberg took up the comparison, already made by Fraunhofer, between some lines in the Sun's spectrum and those in that of sodium gas. By 1861 Kirchhoff could present to the Berlin Academy a chart nearly eight feet long, which appeared to prove the presence in the Sun's atmosphere of many familiar earthly elements. Kirchhoff's work was taken up with tremendous enthusiasm in Italy, France and England — where Sir William Huggins wrote 'Here at last presented itself the very order of work for which in an indefinite way I was looking. . . . A feeling as of inspiration seized me: I felt as if I had it now in my power to lift a veil that had never before been lifted.' Huggins became the first to measure the Doppler effect in stellar spectra, from which the stars' approach towards, or withdrawal from the observer is computed — a matter of great significance in modern cosmology.

From the work of Kirchhoff, Huggins, and many others of the time, there is a direct development to such theories as those constructed independently by C. F. von Weizsäcker and H. A. Bethe in 1938, of the source of solar energy (and as the Sun is considered an average main-sequence star, of stellar energy in general as well).

Today it is accepted as well established that the Sun's power to pour out energy at an almost constant rate through the immense periods of geological time is adequately explained by nuclear physics. From school-days onwards the Sun appears no more mysterious, or worthy of reverence, than a nuclear power station.

For human *knowledge,* there seems to be no relation at all between the physical Sun and the 'aetherial dominations' described by Shelley, who say of themselves:

Our great Republic hears, we are blest, and bless
    or the Bright immortal Lover
That is in the sun and is our Being's sight.

It is a significant fact that Rudolf Steiner was born in 1861, the year of Kirchhoff's chart, which has been described as the 'official opening of the new science of celestial chemistry'. For this meant indeed that mankind crossed a very important threshold, deep into matters that were not indeed 'never known before', but which belonged to holiest mysteries.

The writer of *Les Grands Initiés,* Edouard Schuré, once asked Rudolf Steiner some far-reaching questions about world history, and about his own life, to which Rudolf Steiner gave written answers. These have now been published in part by Madame Rihouet-Coroze in the first volume of her biography of Rudolf Steiner. These indicate that the great leader of esoteric Christianity, Christian Rosenkreuz, looked forward during his work in the Middle Ages to a point in time lying then in the far future. There would come the historic moment when much which had been cultivated through the centuries under conditions of the strictest secrecy would become public. But before the Christian mysteries could be renewed on the open stage of history, certain conditions had to be fulfilled. External scientific knowledge had to reach a provisional solution of certain problems. As the first of these, Christian Rosenkreuz indicated the discovery of spectro-analysis, with its resulting picture of the material constitution of the universe.

We need not stand before the majestic but lifeless

1. THE SUN

shadow-picture of the Sun given by present-day astro-physics feeling *only* tragedy in the fact that mankind has lost an old natural love and honour for the Sun. *Because* this shadow-picture has come about, through much human labour and thought, there can *also* be, freely approachable by all men, a new wisdom of the Sun, a new love and honour enkindled in us. Within the sacramental life of The Christian Community we are given the most direct intimation that we are being led into new mysteries of the Sun by certain Epistles (or seasonal prayers) in the Act of Consecration of Man. Three times during the year the life and being of the Sun are directly described: at Advent, at Easter and during the season of St John, at Midsummer.

The Midsummer Epistle brings before us the Sun as it lived in the experience of humanity in the remote *past:* the original home of Christ, from whence he comes, taking his path to Earth in its ethereal rays. At Advent we hear the Sun as prophet, borne in its splendid chariot, reflected in the rainbow that spans the heavens; the Sun guiding us into the *future.* And at Easter we share in the joy of the Earth's breath, living in the spirit-radiant power of the Sun; man, Earth and Sun as they can be in the immediate *present.*

The words of the Epistles speak to our imagination and to our hearts; the everyday understanding follows very slowly. But it is good if in our thinking we can make tentative bridges between one realm and another — and not leave what we see through the window, what we hear of from the astrophysicist, and what we feel in the Act of Consecration, as things forever apart. Human thought is always tempted to make fixed bound-aries — to regard the human being, for example, as consisting only in what is enclosed by the skin. But such limits are in reality only relative: to every organism belongs much that is spatially outside it — both what is necessary to its life, and its *effects*. In the Indian hymn that inspired Edith Sitwell's poem, the word 'honey', *madhu,* had also the more general meaning of 'effect'. Honey is the 'effect' of bees and belongs essentially to their organism. The lines quoted could be rendered abstractly 'The Sun is the effect of all beings, and all beings are the effect of the Sun. . .' Our abstract and unimaginative thinking makes a good start if it refuses to regard the Sun as only a shining sphere far away, but reckons to the organism of the Sun everything in which its effects live — first the mysterious and mighty realm of the corona, and then the sun-irradiated spheres of all the planets and the Earth. Just as what a man has done in the past — even if it be far away and long ago — remains a part of his being, so the work of photosynthesis in the plant-leaf, for example, should not be regarded as an accidental effect, but an essential revelation of the Sun's being. Man, animal, plant and entire Earth breathe Sun, and through this breath are nourished. The building of carbon into sugars in the plant is a deed of the Sun, to whom carbon is no stranger, and the bees gather as servants of the Sun. Along this path the driest, most middle-aged part of our understanding can be led, and can get younger as it goes.

And in winter, when the work-shops of the Sun are blown away, to serve in the dark soil, and the visible sphere moves on a humbler arc, the rainbow is higher. We are inclined to think of the rainbow as belonging to the whole Earth indifferently; but it has been pointed out by Walter Bühler how particularly it belongs to the *temperate* regions. In the tropics the Sun

1. THE SUN

rises more quickly to a height (above 40°) at which no rainbow can appear; and in arctic regions it is nearly always too cold for rain. Temperate climates in moderately high latitudes (like that of Ireland, for example) are most favourable for rainbows. In these regions Christianity has often proved very much at home.

For the rainbow gives answer according to the nature of its questioner. The first spectro-analysts asked questions about the material constitution of the universe, about distances and temperatures. They were like the young Parsifal, interested only in the weapons and armour of the knights he met, and not at all in their wishes, feelings and character. The rainbow, or the spectrum, is most anxious to speak not about substances, and the shadows they throw, but — through the colours themselves — about the *soul* of the Sun. We have to overcome any idea (though a whole line of dignified philosophers have misled themselves and others into holding it) that colour is only our own subjective, personal experience. This idea seemed justified by the fact that colours do not *belong* to objects in the way that forms do. But in colour we share in the real experiences of light in its encounter with objects:

> And colour breathes that is reflected light.
> The ray and perfume of the Sun is white:
> But when these intermingle as in love
> With earth-bound things, the dream
> begins to move
> (Edith Sitwell, From *A Hymn to Venus*)

We know a person or a being through his relationships; we know the Sun, through the varied joys and sorrows which light has in meeting water and Earth — the colours.

In the richness of this life of soul, we find the promise for the future of man. Could we liken for a moment what is in our own heart and mind, with that fullness and clarity? Should we even call what we presently have a *soul,* compared with this? But man can hope to grow.

While in the realm of eternity the Sun ever gives itself, and is for ever the same, through the course of human history the Sun *changes* — above all by the withdrawal of the Spirit of the Sun and his entry into the Earth. Once Earth 'had her heart in highest heaven' as the *Atharvaveda* says, but now this heart is within her.

When we look at anything — there is change both in ourselves, and in the thing we see; though we forget that something has gone out from ourselves in our gaze. With the Sun, we are aware of the light streaming out — but do not think of anything that is received back. We do not think of 'the eye of heaven' *seeing.* And yet if we reckon the planets to the total organism of the Sun — must not it be influenced in all its members, even in the shining sphere itself, by the history of every part? That earthly beings are part of the Sun's 'honey' is easier to see than that 'This Sun is the honey of all Beings. . . .'

In what sense could this great *cause* of the processes of earthly life be itself an *effect?*

We may perhaps approach an answer if we think of love in the greatest sense in relation to evil. It is possible for love to stream undiminished towards a human being who becomes increasingly evil. Undiminished, but not *unchanged.* Not in the slightest does the love itself take on the nature of evil. The love itself is guiltless; but its mood is deeply influenced, it is an 'effect'. The

1. THE SUN

love really goes out to the one loved; he is in the midst of it, but he cannot *live* in it, unless he understands and accepts how he himself must be seen by the lover.

The Midsummer Epistle speaks of realms into which only the guiltless can enter — and of the word of John the Baptist as conscious of guilt. Man's word can only become sunlike again through the recognition of how far it has fallen. That the thunder of the Sun may not overwhelm him, he needs the repentance John taught, and the mercy brought by Christ.

All the quantities, movements and processes described by the astronomers will in time come to life, like scattered letters gathered into words, and the distinction between a physical world and a moral world will melt away. Then the great distance between Sun and Earth as calculated by the astronomers — 93 000 000 miles (149 000 000 km) — may seem an image of the distance we need, for our slow and uncertain development. And the dark waxing and waning sun-spots, images of the suffering love that beholds human history. And the movement of the Sun will be known as the greatest, most solemn and beautiful of all dances, into which the Earth is drawn.

2. Saturn

In a clear night sky Jupiter and Venus, if they are well above the horizon, draw our attention with a certain insistence. Venus is much brighter than any star; to Jupiter only Sirius among the stars of our skies is comparable. But Saturn is much less conspicuous. It is never as bright as Sirius, though it can at certain times be brighter than any other star we see, and it generally stands out among those in its immediate neighbourhood. But even at opposition, when it shines most strongly, it is often fainter than several stars of the first magnitude. To be sure that we are looking at Saturn, we may have to look at a star map carefully first.

The movements of Saturn in the course of a year follow the same sequence as those of Jupiter, but everything seems slower and more remote. After a fairly long period of invisibility near the Sun, it begins to appear above the eastern horizon not long before sunrise. Gradually the period of visibility becomes longer. After about three months its movement among the fixed stars changes, and becomes a westward one for about four and a half months; in the middle of this retrograde period Saturn is at its brightest, and is visible from evening until morning. In the months that follow Saturn gradually draws nearer to the Sun again, across the night sky, until it becomes invisible in the radiance of sunset. Before this it will have resumed its slow eastward movement amongst the constellations.

As with Jupiter, Saturn's movement through the zodiac brings with it periods when Saturn follows a high course through the sky and those when it is on a low course, like the winter Sun. But with Saturn these periods are longer. During the last twenty years of the twentieth century, in the Northern hemisphere Saturn is descending until it makes its loop in 1988, in the midwinter region of the zodiac, and from then onwards ascending.

Saturn spends some twenty weeks of each year in retrograde movement, falling back as it does so 6½ degrees, while it covers 18 degrees in a period of direct

2. SATURN

movement. The comparable figures for Jupiter are 10 degrees and 43 degrees. Thus while Jupiter from one conjunction to the next advances a little more than a whole sign, Saturn progresses about 12 degrees in a year, and needs two and a half years to complete its passage through a sign, and thirty years for the circuit of the whole zodiac. While the dates of Jupiter's conjunctions and oppositions advance from year to year by a little over a month, Saturn's advance by steps of a fortnight or less.

These rhythms were familiar to astronomers in pre-Christian times, and up to the beginning of the seventeenth century there seemed to be no marked external differences between Saturn and Jupiter except in brightness and relative movement. But when Saturn was first observed through the telescope it was noticed with profound astonishment that its appearance was unlike that of the other planets. Galileo thought at first that he was seeing three bodies close to one another. He wrote in July 1610: 'I have discovered a most extraordinary marvel. The planet Saturn is not one alone, but is composed of three, which almost touch one another and never move nor change with respect to one another.' The first four moons of Jupiter had recently been discovered by him, so he naturally thought he might again be seeing moons, this time very close to the planet. He wrote in another letter: 'So! We have found the court of Jupiter, and two servants for this old man, who help him to walk and never leave his side.' But the observation did not confirm this impression; and it took fifty years to establish that Saturn was surrounded by broad continuous rings, sometimes plainly visible in the telescope, and sometimes vanishing from view. It was Huygens, between 1655 and 1665, who both discovered an actual moon of Saturn and gave an acceptable description of the variable appearance of the rings. He showed that sometimes the northern surface and sometimes the southern surface of the rings is visible, and that there are short intervening periods when the rings are edge-on to the earth. The light from the surface of the rings adds considerably to Saturn's total brilliance; so that it is brightest when the rings are so tilted that the planetary body is as it were embedded within them, and faintest when they present their edge. Within Saturn's thirty-year cycle, there are two such periods of maximum brightness (when Saturn is moving from the constellation of the Scorpion into Sagittarius, and again from the constellation of the Bull into the Twins) and two of minimum brightness (at the end of Leo and between Aquarius and Pisces). This difference is great enough to be noticeable by unaided eyes.

Right up to the present day, the rings of Saturn have been observed with enormous patience and care, and much thought given to their physical composition. It has been observed for example that they are translucent; the light of a star crossed by Saturn has been observed shining through the rings, though with diminished intensity. Many observations have led to the conviction that the rings are made up of a multitude of small separate bodies, possibly ice crystals. For the astronomer, only questions of this kind have seemed real; any feeling for a spiritual significance behind the appearances of Saturn had left him utterly. Such a feeling can still be observed, as an echo, in Galileo's reference to the 'old man'; or, more strongly, in Shakespeare's description of heavy Saturn laughing and leaping with proud-pied April, in Sonnet 98.

2. SATURN

A bridge between the external phenomena of Saturn and its significance for human destiny can be made if we consider the relation of its rhythms to those of an individual life. Saturn returns again to the place where it stood at a man's birth, after making a full circle through the zodiac, only when he is nearly thirty years old. It returns a second time when he is nearly sixty. (This of course disregards Saturn's triple passage through a certain point, if this falls within a loop, in the course of a few months.) We can compare what an interval of twelve years means in a human life, with that of thirty years. Considering the Jupiter rhythm we can speak of a growth in wisdom; with the Saturn rhythm we have to see a radical change in man's cosmic condition, in his standing towards birth and death. Here Time has to be seen not only as teacher, but as the servant of eternity. And indeed for ancient man Saturn was closely identified with the mystery of Time; above all with the tragic sense that Time devours all things, that nothing can long escape the signs of its power. In much of traditional astrology, Saturn was regarded as the bringer of great misfortune, by far the most ominous of all the planets. Not only loss of life, but loss of every kind, outward and inward — of possessions, of servants, of capacities and spirits — is traditionally connected with Saturn.

But when we consult Rudolf Steiner's cosmology, we find something that seems very different. The sphere of Saturn, the whole space encompassed by Saturn's orbit, represents in our present universe a very ancient condition, so to speak three worlds away, in which there was nothing more material than what we encounter in our present world as warmth. Present-day man thinks indeed that warmth can exist only as a quality of some material thing — but this was warmth in itself, warmth as the expression of the sacrificial will of great spiritual beings, the Thrones, who are the companions of the Cherubim and Seraphim in the first hierarchy. Saturn has to do, in the most far-reaching sense, with *beginnings*. How then did it come to be regarded as a planet of misfortune?

On earth we are always trying to establish things in such a way that they will be sure to last; solid, foolproof institutions, for example. We do not succeed; if we did, such institutions would be the greatest barrier to spiritual progress. What seems misfortune is often the breaking in of new possibilities, which our rigid plans would have prevented. Both in great and little things, what seems like loss can often lead us back to the original Divine purposes on which life is really founded.

While Jupiter summons us to more comprehensive and objective thoughts about the world, to a serene and joyful wisdom, Saturn calls for a far-reaching and dedicated power of will. This is the will which is ready to make at the right time the greatest sacrifice.

In the form of a comedy, Shakespeare has wonderfully expressed all this in *The Merchant of Venice*. The people of the play are of course real human beings, not examples of principles; but as real human beings do, they play cosmic parts. Shylock bears traditional saturnine marks; he is melancholic, obstinate, miserly, haunted by fear for his possessions. But the true heavenly Saturn is chosen by Bassanio, when he asks for the leaden casket. (Lead, as Shakespeare knows well, is Saturn's metal.) This casket bears the words 'Who chooses me, must give and hazard all he hath'. Bassanio has reason to feel the significance of this; for Antonio,

2. SATURN

out of his great friendship, has risked everything for his sake. In one of Shakespeare's sources for the play the leaden vessels bears the words 'Who so chooseth me, shall find that God hath disposed for him'. The deepest will which a man can find in himself, for his own action, is at the same time God's will for him. Antonio's action, which brings him so near to death, is the source of new life for Bassanio. And even Antonio's decree, at the very end, that Shylock should become a Christian — which seems so absurd to a modern audience — really means that he wishes to share with Shylock the most precious thing he has.

If Christianity had not come to mankind, it would have become more and more difficult for human souls to participate truly and freely in the gifts of Saturn. Rudolf Steiner describes that in the very ancient world men recognized three kinds of birth. There was the Moon-birth at the beginning of physical life. This was followed at the age of thirty by the Sun-birth at which men felt a strong individual impulse entering their being. And at death there was the Saturn-birth. But by the time of the mystery of Golgotha, a sorrowful feeling was widespread among men of insight, that unless a man had passed through initiation in a school of the mysteries during his earthly life, he would only be a troubled wanderer among shadows after death. Christ came not only to help the living, but also the dead.

At the time of Christ, the visible constellations in the sky corresponded on the whole with the signs, the equal divisions of the sun's path. At the time of the Passion, Saturn stood in the sign and constellation of Cancer; in deeply hidden inwardness, one age of the universe came to an end, and another began. (The symbol of Cancer, inward-going and outward-going spirals, expresses this.)

But the constellation Cancer in the sky is a short one. And about the time of the first Whitsun Saturn passed into the constellation of the Lion, where it was later joined by Jupiter. In the disciples, and before long in Paul too, quite a new relationship to death was developing. They knew that through the power of Christ the star of their true being, their god-given individuality, would shine out most clearly at the moment of martyrdom; and this fiery enthusiasm was to endure through the first three centuries of Christendom.

In childhood we learn to stand upright, and to walk; and it is really Christ's power which accomplishes this in us. In the life after death we have to learn how to stand in the spiritual world through the influences of Saturn. Christ gives to his disciples on Earth the deep confidence that they are already walking with him in the spirit, as they walk on earth. In our time this confidence has to be sought afresh; the will to find it has to awaken first in the stubborn realms of our heads.

At the beginning of the age of the Archangel Michael (according both to Rudolf Steiner and the medieval traditions passed on by Trithemius of Sponheim) in 1879 Saturn stood in the signs of Aries, the Ram. From Aries there stream those forces which so form the human head that man in his whole body can be an upright being. When the Michael age began, thoughts were beginning to take possession of mankind which would obliterate more and more the significance of this uprightness, identifying the nature of man with that of the animals.

In 1996, for the fourth time since the beginning of the Michael age, Saturn will stand in Aries at the spring-oint of the Sun; more urgently than ever, humanity will be called upon to give a new direction to the whole of earthly civilization, awakening its spiritual Will.

3. Jupiter

When we think about the destiny of an individual human being, we have generally in mind the main events in his human relationships, in his health and sickness, and in his work. But among the many elements in a human destiny the natural phenomena a man is able to see are significant too. Not only the more striking things — earthquakes, volcanic eruptions, great comets — but also the quiet, regular features of our environment matter in our lives. What kinds of animal are familiar to us at this or that stage of our lives, what trees there are in our environment, whether we live in a climate and region where rainbows frequent or not — all this is part of our destiny, part of all that the heavenly powers have intended to show us.

When we look up from the earth, the stars and planets are seen differently at different places and times. Some heavenly events are clearly visible, some are hidden in various ways from our senses. How and when some fact of the heavens becomes known to humanity or to a particular person can indicate much to us, both about the world and about man.

We have, for example, among the planets striking differences in their visibility and their motions. It is very easy to see Jupiter, and relatively easy to understand and remember those of its movements which our eyes can see. It is of course the habit of our time to think of such movement as merely apparent — but we can come to understand that it is important to remember appearances, and not to substitute for them an imagined model which we suppose to be more real. (It is all the more natural to do this, of course, because the observations most of us succeed in making, living in towns and cloudy climates, are so very fragmentary.)

Among the planets, only Venus and for brief periods Mars can be brighter than Jupiter at its brightest. But while Venus is always an evening or a morning star, never shining all night through, and Mars can culminate at midnight only alternate years, Jupiter has its time of all-night splendour each year, and is bright in the sky for months before and after this time. Only for about six weeks in each year is Jupiter invisible for us through its nearness to the Sun. Yet, like the Sun, Jupiter sometimes follows a high path, sometimes a low path across the sky, and is therefore above the horizon for a longer or shorter time from its rising to its setting like the summer or winter Sun.

Jupiter is on an ascending path through the zodiac from February 1984 to August 1989, reaching the point of the spring equinox (when it rises and sets due east and west) at the beginning of March, 1987. From 1989 it has a descending course, crossing the point at which the Sun stands at the autumn equinox (September 23) early in October 1992, reaching the lowest point at the beginning of 1996. So at the end of the century it will be close to the spring point on its ascending course. (This description of course applies to the northern hemisphere.)

A period of Jupiter's visibility always begins with its emergence as a morning star, its 'heliacal rising', when it rises in the east not long before sunrise. Each morning that follows it rises a few minutes earlier in relation to the sun, until after about ten weeks it rises already at midnight.

3. JUPITER

It is striking to compare the changing relation of an 'outer' planet like Jupiter to the Sun with that of the Moon. We see the young Moon becoming visible not out of the sunrise but out of the sunset, rapidly becoming later each evening in its setting, with a wider extent of sky between it and the Sun. And when, after nearly a fortnight, the Moon is full, it rises about sunset and culminates at midnight. The time of Jupiter's greatest brightness, which has been described above, corresponds to Full Moon; it is an 'opposition' to the Sun. But it has been reached on the other path — from the sunrise, instead of from the sunset. And there is another less obvious difference. The Moon is always proceeding through the constellations which surround it in the direction from west to east, as is the Sun. And when Jupiter becomes visible as morning star, it is doing this too; though this movement is very slow, a star which was just beside it will after a month be about six or seven degrees to the west of it. (The visible discs of the Sun and the Full Moon are both very close to half a degree in diameter.) But about four months after it has become visible, the motion of Jupiter among the stars changes. Having become very slow, and then stationary, it begins to move westwards among the stars, in the same direction as their swift nightly movement. It continues in this direction for four months (covering altogether about ten degrees); and it is in the middle of this period that 'full' Jupiter, its opposition to the Sun, occurs.

From ancient times the retrograde movements of the planets — plainly visible with Mars, Jupiter and Saturn — have presented a far-reaching problem to astronomers. In Greek and Roman times astronomers often saw their task as that of reconciling what appears as an

27

irregular movement with the dignity of the heavens, which seemed to demand movements in perfect circles. The Ptolemaic system went far in satisfying this demand. From the sixteenth century onwards the impulse arose to explain planetary movement in terms of earthly physics; and immense labours were necessary, above all by Kepler (using the observations of Tycho Brahe) and Newton, before the Copernican system could be developed in a way that seemed to fulfil this task. And in the future still other pictures of the relationship of movement within the solar system will be needed, much more closely related to the experience and the physiology of man.

We can take a significant step towards a qualitative experience of the planetary movements if we feel how this third of the year, in which Jupiter is making its loop among the stars, is related to the seasons of our earthly year. Each year the whole process — becoming retrograde, reaching opposition, assuming 'direct' (eastward) movement again — is repeated, but falls just over a month later in the year. Thus if in one year Jupiter has been at its brightest in midwinter, three years later this brightest period will be in the spring, and after another three years, at midsummer. After twelve years the whole process will have come back to almost the same points in the years as at the beginning.

In general each succeeding opposition of Jupiter will fall in a new constellation of the zodiac. (This is not *precisely* so, because in each twelve-year period there are eleven oppositions, and the constellations are of different lengths). These positions of Jupiter can be regarded, as has long been done in the east, as related to the distinctive qualities of the year in which they fall.

When a child is born, Jupiter will be moving on a

3. JUPITER

high or low arch, and will be near the Sun or far from it. Near the twelfth, twenty-fourth and thirty-sixth birthdays, and so on through life, these relationships will be repeated. And we can make a bridge from the outward phenomena of Jupiter to its significance for the life of the soul if we ask: can we recognize this rhythm in any characteristic way in a human life? We must be prepared to find that such things are very much hidden beneath the surface of events, and by no means easy to recognize when we are close to them. But if anyone looks back on points in his life separated by a twelve-year interval, it is quite plain that these mark a considerable development of his *consciousness*. Unless we have been very much asleep mentally, we have a much richer and more objective picture of the world after each twelve-year period than we had at its beginning. It might well be said, of course, that this applies to shorter periods too. But we can often see that over shorter periods a man may go through upheavals in his ideas which do not look at all like a development — and yet in a longer period all this may be included in a movement towards objectivity, towards wisdom. It is into all that can be described as wisdom, in the most genuine and far-reaching senses of this word, that Jupiter seeks to lead us; and to feel this is in accordance both with ancient traditions and with the renewal of spiritual understanding for the heavens which is coming about in our time.

We have not in our time a secure conception of what wisdom is. We speak with some assurance about 'ability' or 'brilliance' — but it is not easy to think of any public figures, artists, or writers of our time who would be described with complete conviction as 'wise'. We are more inclined to attribute courage and independence to the old people we know than wisdom. And yet there are moments when we seem to encounter in all sorts of people, at all ages, and even in ourselves, a calm light of understanding, bringing order into the complexities of life.

Jupiter in the sky may nevertheless seem very far away, and we may be tempted to feel that his wisdom is cold and detached from the immediate concerns of our life. But the quality of Jupiter is not really like this, nor was it so understood in the past; we need only think of the word 'jovial' to see that philosophy and cheerfulness were not always considered incompatible.

Rudolf Steiner has described a very significant relationship of the sphere of Jupiter to the human soul, which is established a considerable time before the soul enters through birth into earthly existence. In far realms of the spirit, the soul resolves to assume once again the weight of an earthly body. It begins to long for the conditions of earth, with all their hindrances, which provide man with the opportunity for developing his inner freedom. And in the sphere of Jupiter there streams into the soul a joyful sense that it will be able to encounter the heaviness of earth effectively. Joy in the acceptance and fulfilment of the tasks which are brought to us by our earthly destiny is a manifestation of the realm of Jupiter.

When in spring we see some of the earliest flowers with clear light yellow petals, this mood of Jupiter is reflected to us from the banks and meadows. To see many primroses, cowslips, and daffodils, in surroundings where they grow freely, is a good gift of destiny, akin to the conscious sight of Jupiter in the fields of the stars.

Where Christianity has been rightly understood a

3. JUPITER

joyous acceptance of the tasks of earth has always been present. Earth has been consecrated by Christ's resolve to undertake its burdens. In the life of Jesus, the rhythm of Jupiter is indicated by St Luke when he describes how the twelve-year-old Jesus remains behind in the Temple to speak with the learned men, and astonishes them by his wisdom. When this happened Jupiter was near opposition, shining at its brightest through the whole night in the region of Virgo and Libra. And in the years from the beginning of John the Baptist's proclamation of Christ's coming, up to the conversion of St Paul, from about AD 28 to 35, Jupiter took the path from Aquarius to Leo; from the transforming water of baptism to the glowing fire of that unconquerable enthusiasm which the vision of Christ inspired in Paul.

At the time of the mystery of Golgotha, Jupiter was an evening star, between Mars and Saturn, while Venus and Mercury were morning stars. It belongs to the environment of Holy Week and Easter that after sunset Jupiter was shining in the south-west not far east of the red glow of Mars; both planets were in the region of Gemini, the Twins. So will it have shone, if the evening was clear, when the two disciples returned swiftly from Emmaus to Jerusalem, bearing their account of the companion who had interpreted to them the wisdom of Moses and the prophets. For those familiar with Greek mythology, as St Luke was, the Twins bore the names of the immortal and the mortal brothers, whom death should not separate. The first Christians were to proclaim that the true immortal brother had indeed come, and had suffered death, that he might not be separated from man. This would be the heart of their wisdom, which would seem folly both to Jew and gentile.

In our time too the wisdom which begins to understand the Christ in his significance for the universe and for nature is rejected as folly. This should not make us restless or uncertain. For any real truth needs time to establish itself in us; superficial or mistaken impressions come and go quickly. And even ancient and heavy errors will be rolled away at last, as the great stone was rolled from the entrance to the sepulchre.

4. Mars

In the first part of *King Henry VI,* Charles the Dauphin says,

>Mars his true moving, even as in the heavens
>So in the earth, to this day is not known.
>Late did he shine upon the English side;
>Now we are victors; upon us he smiles.

When the play was written, about 1591, Tycho Brahe was still making his long series of precise observations of Mars, on which Kepler could base his great laws of the planetary system. Of all the planets, Mars shows the greatest variability of movement; for a picture of the solar system based on movement in perfect circles, these provided the greatest difficulties.

Like Saturn and Jupiter, Mars is at its brightest when it is in opposition to the Sun. This happens when it is in retrograde movement, that is to say going slowly westwards against the background of the fixed stars. It does this for a period of from two to nearly three months, as compared with Jupiter's four months and Saturn's over four-and-a-half. During this backward

4. MARS

movement Mars covers on an average half a zodiacal constellation. But when it begins to move forwards (eastwards) again, it moves very much more swiftly than Jupiter or Saturn. Half a year or more, including the retrograde period, may have been spent in one constellation; but in the next eight months or so Mars will cover half the entire zodiac. Mars has thus a much longer period as an evening star than Jupiter or Saturn. From one point of view, we can say that it falls back into the sunset much more slowly; or, from the point of view of the starry background, that it moves so swiftly that the Sun has great trouble in catching it up.

Thus while Saturn and Jupiter have a period of greatest visibility *and* a period of invisibility within each year, these fall for Mars in successive years.

From one opposition to the next takes, however, more than two years, by an amount that varies from five to eleven weeks. In the course of fifteen years, there are seven oppositions, and the dates of successive opposition have advanced by irregular hops through a whole year. During this period Mars will have marked out, as it were, seven points on the zodiac, the places among the stars when it shone most brightly within each opposition year. But these points vary considerably among themselves in brightness. The reddish glow of Mars is most brilliant when it stands at opposition in the constellation of Aquarius. It is then considerably brighter than Jupiter.

The phenomena connected with Mars are rich in contrast. In relation to the stars it moves very much more quickly than the other outer planets, except when it is making a loop. But its relation to the Sun changes much more slowly; when it begins to appear before the sunrise, or draws near to the sunset, the daily increase or decrease of angle is in comparison very slight. And at the times when it is seen only briefly, as morning or evening star, it is also very insignificant in appearance; for considerable periods it is fainter than Saturn. And even at its brightest there is in the red glow a suggestion of darkness; red appears, as Goethe shows in his work on colour, when light has to reach us after a heavy struggle with the dark.

Thus what can be seen and remembered about the external appearance of Mars does suggest quite strongly what is described of its spiritual aspect, both traditionally and in our own time by Rudolf Steiner. Mars has to do with conflict; both with physical warfare, as in the words of the Dauphin, and with conflicts of mind and soul. Even the period which Mars takes to go round the entire zodiac, nearly two years, can tell us something about this. For while Saturn's thirty-year period brings changes in our relationship to birth and death, and Jupiter's twelve-year period ripens us in wisdom, the measure Mars gives us applies well to many kinds of human conflict. Wars cannot really go on consistently for thirty or for a hundred years; wars so described have really proceeded in bursts of activity and periods of quiescence, very much changing in character, and even their participants, during such a period. But within two, or four, or six years — what limitless grief and human loss wars brought! We might easily look up at Mars and regard it as the planet concerned with the greatest misfortune of mankind.

But this would really be only a one-sided and superficial conception, both of history and of the planet. The experience of conflict has been one of the greatest teachers. It has taught courage, endurance, and self-sacrifice — even compassion and hope. In some way

34

4. MARS

or other it has always stirred human beings to look for something that would transcend their differences and rejoin them in unity. The real genius of Mars leads in this direction — towards the complete experience of extremes, and their inclusion in a great synthesis.

This genius is at work in human speech. Almost every word brings together contrasting elements; we respond, though we hardly notice it, to the radically different qualities of vowels and consonants following each other in sequence. The interplay of similarity and difference continues into the sentence, and permeates every structure built up of words. It shows itself most evidently in poetry, where the contrasts of Mars are tempered by the gentle beauty of Venus.

But the reconciling power which really lies in the very nature of speech is not easily used. Words indeed come handy as weapons; and even where there is some willingness to meet in discussion, there can be great difficulty in finding a common language. Much experience of this has been gathered by the representatives of different Christian churches in their meeting with each other; and even within a single family vast abysses of misunderstanding can open up. And yet — speech is ultimately an instrument for peace.

When we come into earthly life through the door of birth, we bring with us effects of our experience, extending over long periods of our existence before birth, in the regions of the spiritual world. The visible planets are like signposts, pointing towards realms which are the dwelling places of many kinds of spiritual beings, from whom we have received powers for use on earth. For thousands of years, the souls of men have brought courage to overcome the difficulties of Earth from their sojourn in the sphere of Mars. But in recent centuries a particular quality has entered more and more into this courage.

Wherever we may live on earth, we are surrounded by a climate of opinion which has a strong effect. There are views which those in our environment find acceptable, and others which shock or disturb them, or which would seem to them so unusual that they could hardly imagine them seriously held by anyone. At the present time, for instance, most people absorb from scientific or historical textbooks at school statements which they regard as having absolute validity, which it would be crazy to contradict. There are other opinions which are more a question of something like fashion; they may last only for a few years, but are for the time widely believed and obeyed. As an extreme case we may take the ideas imposed upon whole populations by communist or nationalist dictatorships.

However pervasive the influence of a widely held opinion may be, or however strictly it may be imposed, we can find within us the power to question it. Just as we can make ourselves free of our own past beliefs, we can look afresh at any matter about which we have listened to statements from others, and try to hammer out a new judgment for ourselves. To do this — even if it is our intention to keep quiet about our unorthodox opinion — requires courage. It is more comfortable to go on in an accepted opinion, whether our own or someone else's. It is the courage to look afresh and for ourselves, which the noblest spirits of the Mars sphere seek to give us before we are born.

It can often appear as if human beings, each seeking to form their own individual opinions, were destined to grow more and more different, and less and less capable of understanding each other. But the process of

4. MARS

liberating ourselves from our own past opinions, and from the pressure of external public opinion, is really one that helps us to enter into the true judgments of others who have achieved them for themselves. This depends indeed upon our willingness to use the inner freedom and mobility we are acquiring positively towards others; it depends upon the development of some warmth and confidence towards the other person, which are not always easy for us to produce.

When we look at conflicts in which we are not ourselves involved — or which happen before us on the stage, or which we read about — it is natural to identify ourselves with one side or the other. Whether the struggle is between Greeks and Trojans three thousand years ago, or between football teams this Saturday afternoon, there can be a good deal of pleasure in this. On the other hand, we can look on personal conflicts of which we know, or on present-day wars, in a mood of despair. It can seem that there is no real solution, and that any apparent victory for either side is really only a disaster for all.

But it is not necessary, when we look at a conflict, either to identify ourselves with one side, or to despair. Long ago, Krishna taught Arjuna about this, as the *Bhagavadgita* describes. We may have to participate in one way or another in an earthly battle; but we can also feel ourselves part of a great process, through which all divisions are transcended and healed. Into an earth in which Mars forces had come to strong expression, there enter gently the healing influences of Mercury. We can turn towards Mars with the faith that each difference can gradually be carried on to a level where it is no longer destructive; where it can be as positively creative as the mingling of sounds in right speech.

It would indeed be an illusion to suppose that our capacity for destructive conflict is nearly exhausted. War has become an appalling anachronism, but a toughly persistent one. We shall only begin to leave it behind in so far as we can raise our vision to the great war in heaven, to Michael as he casts down the dragon, and keeps him beneath his feet. When man in the contemplation of Michael grows strong enough inwardly, that which corresponds to the dragon within him can no longer darken his heart; then the Mars in him will cease to be a power that wounds. In the sword of Michael iron is no longer destructive, but a power that guards the holiest from profanation.

In the Apocalypse John describes his vision of the woman clothed with the Sun, with the Moon under her feet, and on her head a crown of twelve stars. She is persecuted by the dragon whom Michael casts down. The dragon seeks to devour her child, who is to shepherd all nations with a staff of iron.

In the starry heavens, we find this woman as the constellation Virgo. We have not yet learnt to purify the iron within us, to make it virgin. The tragic world war of the twentieth century, which has never really ended, but has continued since 1914 in changing forms, broke out as Mars entered the constellation of the Virgin; it is a terrible rejection of the starry guidance. Before the end of the century, there will be great need of all that we can do to transform and to redeem our relationship to Mars.

We may be very much helped in this if we can begin to feel the significance of a statement by Rudolf Steiner about the spiritual sphere of Mars, that since the beginning of the seventeenth century this sphere itself has begun to be transformed by a great spiritual deed. Since then the peace-bringing, healing power of that being who once lived in Gautama Buddha has worked in and from Mars, implanting strength and inwardness into countless souls, and will continue to do so for ages to come.

5. Mercury

When the sky above the eastern or western horizon is very clear, in countries with temperate climates similar to that of Northern Europe Mercury can sometimes be seen for a short time before sunrise or after sunset. But it is only during a few weeks in each year that there is this opportunity. It has been reckoned that in Central Europe Mercury could be visible only for eighteen to twenty hours in a year, adding together each of the short periods in the morning and in the evening. The actual visibility for any place in Britain, with its misty horizon, is very much less. Many people interested in the stars, but without easy access to a telescope, have never seen Mercury. In the United States and the populated parts of Canada, visibility is better than in Britain. In both hemispheres the most promising times are when Mercury is an evening star in respective spring, or a morning star in the autumn.

The pattern of the movements of Mercury is like that of Venus, but it is much more rapid and elusive. In dry countries nearer to the equator, conditions for the observation of Mercury are very much more favourable; nevertheless it is a relatively small part of its path that can be seen without telescopes, because it moves in such close attendance on the Sun. Elisabeth Vreede, the first head of the Section for Astronomy at the Goetheanum, has indeed described the relationship of Mercury to the Sun as comparable to that of the Moon to the Earth. Both with Venus and Mercury phases similar to those of the Moon can be observed through the telescope. But Dr Vreede's comparison is not just an external one; superficially, it could be made for all the planets (though in size Mercury approaches the smallness of our Moon). Mercury in its rhythm — and the rhythms always help to lead towards an inner consideration of the planets — appears as a mediator between Sun periods and Moon periods. These rhythms have a great variability; with all the planets, the 'synodic' periods, the length of time from one conjunction with the Sun to the next, are to some extent irregular, but Mercury's variation is the greatest in relation to its period.

Like Venus, Mercury has two kinds of conjunction with the Sun: 'superior', when it lies beyond the Sun, and 'inferior', when it is between the Sun and the Earth. At inferior conjunction the tiny disc of Mercury can sometimes be seen (with instruments) crossing the disc of the Sun. Such 'transits' also occur with Venus, but they are very rare; there are none at all in the twentieth century. Transits of Mercury occur at irregular intervals; in a century there are thirteen or fourteen such transits. The first astronomer who was able to foretell transits of Mercury and Venus was Kepler. At the end of his life he called his fellow astronomers to observe the rare event in 1631, that transits of Mercury and Venus would occur within a few weeks of one another. But he did not live to observe these himself from the Earth.

5. MERCURY

The average period from one inferior conjunction of Mercury to the next is nearly 116 days, a few days short of four months. Thus there are usually three inferior conjunctions, and between them three superior conjunctions each year. Marked out on the circle of the zodiac these make a wonderful six-pointed star, just as the conjunctions of Venus over four years make the five-pointed star, the pentagram. (If we take the points of a pentagram in the sequence in which they can be traced by a continuous line — as when a eurythmist steps a pentagram — then Venus requires eight years to trace the form, taking either superior or inferior conjunctions separately, as is described in the following chapter.)

Three synodic periods of Mercury, three times 116 days, make 348 days, seventeen days short of a year. Thus the conjunctions of Mercury will fall behind their predecessors of the previous year; this falling behind, taking into account the variations in the period itself, will usually amount to two or three weeks. Thus while the 'star' formed by the Venus conjunctions moves only very slowly in the zodiac, that of Mercury changes very rapidly; one point of the star will generally move out of a constellation of the zodiac after two years.

Thus the character of Mercury's relationship to Sun and Earth is expressed by a rhythm of *threes*; the Mercury-period is such that it nearly extends to *four* Moon-periods of twenty-nine and a half days. The Moon meets Mercury twice as evening star, soon after New Moon, and twice as morning star, shortly before New Moon, in each Mercury period. In this relationship we can find something expressed of the great traditional task of Mercury as mediator between the forces of Sun, Moon and Earth.

Outwardly little visible, engaged in continual

transformation, close always to the thresholds of day and night — these are some of the qualities which lead us from the study of the astronomical Mercury towards an understanding of its spiritual reality. Ancient tradition connects Mercury with skill and agility, even the skill required for stealing. But the skill through which we can see most deeply into the nature of Mercury is that of the healer. We still find the staff of Mercury used as a symbol of medicine, though its significance has long been forgotten. The true healer will always help his patient to find a new balance between waking and sleeping, between activity and rest, between body, soul and spirit. The purposes of the spirit are hindered when the soul is unable to carry them over harmoniously into expression by the body. Every illness is a trial of the soul, a stage on its journey. Mercury is the helper on the journey, the interpreter of the language of destiny.

In his lectures on the Gospel of St Matthew, Rudolf Steiner pointed to significant moments in the history of the people of Israel. He described the whole course of their destiny from the time of Moses to the Incarnation of Christ as a spiritual journey from the Moon to the Sun. In this way the people of Israel received impulses which came from the realms of Mercury and Venus. The entry of the Mercurial impulse was marked by the rule of David, of the Venus impulse by the Babylonian captivity.

What came through Moses as the great revelation of the Law had the effect, as St Paul describes in the Epistle to the Romans, of showing with ruthless clarity how men had fallen away from the divine will. No one could keep the Law; and the sin of those who knew the Law, but disobeyed it, was all the deeper. The Law itself could warn, but not heal.

5. MERCURY

Moses indeed looked upon Christ and could bring some of his healing power to Israel during the forty years in the wilderness. But it was through David that a way of understanding for the relationship between God and man could be unfolded, which began to bring liberation from the rigidity of the Law and show how the divine healing power works. In the Psalms of David an imaginative language is spoken which overcomes the barriers between the earthly and the heavenly. As Dr Rudolf Frieling has shown so abundantly in his book, *Hidden Treasure in the Psalms*, this language contains great wisdom, and yet this is expressed with a delicate poetic simplicity. David is himself shepherd, musician and king. As ruler he can still dance before the Lord. His devotion is joyful as well as earnest. Through this mercurial quality he can find the right pictures, through which the gentleness and mercy of Christ can be described. 'The Lord is my shepherd, I shall not want'.

The whole of the twenty-third Psalm is a great healing imagination, bringing comfort to the soul in the trials of life and in the valley of the shadow of death. And the influence of David's vision of Christ worked on in the centuries that followed him, inspiring other psalmists, and being renewed and developed by the prophets.

In the New Testament the same quality is to be found in the Gospel of St Luke. Here are to be found the most wonderful pictures of mercy and healing, for example the parables of the Lost Sheep, the Prodigal Son, and the Good Samaritan. And here it is described how the shepherds come to 'David's city', Bethlehem.

St Luke was a physician with a deep and poetic understanding both for the Old Testament and for the language and literature of Greece. He accompanied

Paul on some of the most significant journeys. Like Paul, he wished to bring what had hardened in the traditions of the Jewish people into life and movement again. In the Acts of the Apostles, he describes for us with profound compassion and understanding the trial and martyrdom of Stephen, and later on Paul's dealings with the Apostles at Jerusalem, through which it becomes accepted that his new relationship to the Law will become general among Gentile converts to Christianity. And Luke describes Paul's stormy journeys as prisoner from Jerusalem to Rome.

The Romans rightly felt themselves as quite particularly the people of Mars; warlike, courageous, and rigid, they produced many dominating personalities, and even the women of Rome have often a Martian austerity and purposefulness. Paul himself was proud to be a Roman citizen; but he came to Rome to show that the Mars-age of the Earth was ending, and should give way to the age of Christ the healer. During all future ages of the Earth, the fallen Mars-forces are to be purified and redeemed.

Christ works indeed through the holiest powers of all the planets; he has the seven stars in his right hand. And he is rightly seen as Sun, the light of the world. Nevertheless the Christ does work in a particular way for the future of the Earth through the forces of Mercury. For the ancient world Mercury as evening star was the leader of the souls of the dead, bringing them to their encounter with the holy power of truth. Now it is Christ himself who awakens and leads the souls for their journeys into spiritual heights. He has become the great interpreter of all that is strange to us upon this journey.

For the medieval alchemists Mercury was not only at

5. MERCURY

work in the metal that bears its name — but in every drop of water, in all the processes of fluidity. When in sacramental and ritual acts, for example in Christian baptism generally, and in some forms of funeral service, including those of The Christian Community, water is used, the significance of the forces of Mercury can be felt — gently helping the entry into earthly life, and giving strength to our will when we set out on the great journey towards the spiritual sphere of the Sun, and beyond it, after death.

6. Venus

There is a clear sky after sunset. It is still bright in the west, and suddenly we notice with astonishment, a little to the left of where the Sun has gone down, and not very high in the sky, a star already shining. Many, many people, who otherwise take little interest in the stars, may well have noticed Venus as evening star under such conditions. Or they may have observed in a sky already dark, in which many stars have appeared, Venus blazing near the western horizon, far brighter than any other star or planet. And many, looking at Venus, may well have had rather a special feeling — as if they here encountered something less remote than other heavenly events, something that calls forth a deep restlessness and longing.

The movements of Venus as experienced from the earth are radically different from those of the 'outer' planets, Mars, Jupiter and Saturn. These we can see high in the southern sky at midnight, when they are near opposition to the Sun, and they are then indeed at their brightest. But Venus is never to be seen in this position; if it is visible as much as four hours after sunset, or four hours before sunrise, it will then be low on the western or eastern horizon. We may try to follow the movements of Venus from the moment that it becomes visible as a morning star in the east. It is then seen only for a few minutes before disappearing in the morning sunlight. But each morning it appears sooner, and is brighter, until after about five weeks it rises some three hours before the Sun, and shines with greatest brilliance. For another five weeks it continues to rise earlier in relation to the Sun, up to about four hours, but is not quite so bright. Then the time between its rising and that of the Sun begins to diminish slowly. For six months it grows fainter, and nearer to the sunrise, until it is no longer visible as morning star. Then for about three months it is not to be seen.

Now the opposite process begins. At first very faintly and briefly, Venus is visible in the west after sunset. Very slowly, over six months, its distance from the setting sun increases. When the angle between the Sun and Venus is at its greatest, at the end of this six months, it is a little more than a quarter of the Sun's arch across the sky. Now the distance between Sun and Venus diminishes, but Venus continues to grow brighter, reaching its greatest brilliance after five weeks. Then after another five weeks, the period of visibility after sunset having become very short, it again disappears. But this time not for so long; the new period as morning star will begin some three weeks later.

If we attempt in this way to picture the sequence of events as they can be seen with ordinary eyesight,

6. VENUS

without being concerned for the moment about any explanation for them, a striking contrast can become apparent. When Venus first appears as morning star, it does so like the outer planets, detaching itself from the sunrise. But when it appears as evening star, it does so like the Moon, detaching itself from the sunset, though of course much more slowly. But in the later part of its period as morning star, Venus behaves like the old Moon, disappearing into the sunrise. And in the later part of its period as evening star, Venus is like the outer planets, disappearing into the sunset.

While the outer planets come into opposition to the Sun during their retrograde periods, when they are at their brightest, Venus and Mercury are never in opposition to the Sun, as should be clear from the description of the movements of Venus that has been given. Though they too have retrograde periods, these are much less conspicuous. Venus is retrograde for about six weeks and part of this movement falls in the period of invisibility as Venus changes from evening star to morning star. The length of this movement is about half a sign; it is in space the longest retrograde movement of any planet, except sometimes that of Mars, and the shortest in time, except for that of Mercury.

So far the movements of Venus have been described simply as phenomena without trying to picture the changing spatial relationships of Venus, Sun and Earth in depth. Up to the sixteenth century, the prevailing picture of the movement of Venus was that it performed a circle which lay in its entirety between the Earth and the Sun — so that in the complete period Venus was pictured as passing twice between the Earth and the Sun (though not of course directly across the Sun's disc). But in the Copernican picture, Venus has a smaller orbit round the Sun than that of the Earth, and thus passes in each period once in front of the Sun and once behind it. With the use of the telescope this received a striking and beautiful confirmation. In 1600 Galileo observed through his telescope that Venus shows phases like those of the Moon. Before he published his discovery, he sent an enigmatic Latin sentence to Kepler, the letters of which could be rearranged to mean 'The Mother of the Loves imitates the phases of Cynthia'. Seen through the telescope, Venus as evening star comes to resemble more and more the fine sickle of the young Moon. But while the Moon always appears to us nearly the same size, the disc of Venus grows more than six times as great from the beginning to the end of its nine months as evening star. Thus Galileo could observe that Venus as evening star is approaching the Earth. From the small Full Venus it wanes (and grows!) to a Half Venus at greatest eastern elongation, and is at its brightest as a sickle, resembling the three-day-old Moon. Similarly as morning star Venus begins as a great but narrow sickle, turned towards the rising Sun like that of the waning Moon, and grows smaller as more of its disc is illuminated.

'Full' Venus is thus the moment when the planet is furthest from the Earth, beyond the Sun; this is called 'superior conjunction', which falls during the longer period of invisibility. 'New' Venus comes when the planet is nearest to the earth, between us and the Sun, during the shorter period of invisibility; this is called 'inferior conjunction'. At superior conjunction Venus passes the Sun eastwards, becoming an evening star at its faintest. At inferior conjunction Venus passes the Sun westwards to become a morning star, soon to be at its brightest.

6. VENUS

In the description of the outer planets it has been shown how their oppositions in their sequence make a characteristic pattern; Saturn and Jupiter reaching opposition each year, Mars alternate years. Venus completes its whole 'synodic period', from one superior conjunction to the next, in about nineteen and a half months (584 days). (This is not 'the Venus year', its sidereal period — in which Venus reaches the same point in the zodiac as seen from the Sun — which is 225 days.)

In the very detailed account of the visible movements of the Sun, Moon and planets by Joachim Schultz, from which much information for these studies has been drawn, it is shown what a wonderful pattern is traced out through the years by the conjunctions of Venus. These positions make in the zodiac an almost perfect five-pointed star, a pentagram, in that order in which we would have to make a pentagram if we were drawing or running it in a continuous line, from alternate point to alternate point. (Those who have done this form in eurythmy will remember how deeply satisfying it is, moved by five people at once.) At the end of eight years the point of the pentagram is less than three degrees from where it began, so that this great star in the zodiac moves only very slowly; each point of the pentagram takes about a century to move through one constellation. The inferior conjunctions make in their sequence a precisely similar pentagram, each inferior conjunction falling at almost the same point in the zodiac as the superior conjunctions four years earlier and four years later.

· While we on earth try to reduce everything to neatly commensurable ratios, generally decimal or dual, it is impressive to consider how in the great universe Jupiter divides the zodiac by twelve or by eleven, and Venus by five — always with small differences through which their mutual relationships continually alter.

It is not only in the great universe that we can hope to find such an interplay of rhythms. In recent years a great deal of scientific study has been given to the time rhythms in plants and animals — connected as they plainly are with the course of the year, and that of the day. These often show both a precision and a variability from which there is much to learn. Researchers inspired by Rudolf Steiner have gone further, and have taken the first tentative steps towards following the traces of planetary rhythms in the world of life, and in particular in the growth of plants. Joachim Schultz, who died in 1953, published in 1949 a remarkable article showing how the spirals which are characteristic of plant growth reflect the relationship of the earth to the planets. For some time careful study has been given to the angles at which leaves follow one another upon a stem. The plant families vary in this, though there is an angle which can be shown to be fundamental, that of the Golden Ratio. But in their characteristic variations Joachim Schultz is able to relate great groups of plants to the movements of the planets, as they approach or withdraw from the earth. Many dicotyledons show the angle 144°, two-fifths, but Schultz takes as the best example of this the widespread rose family, the Rosaceae. This family includes most of the familiar fruit trees and shrubs, the apple, pear, plum, peach, blackberry and strawberry; also the mountain ash and the hawthorn, and some small herbs, for example, *Alchemilla*. Here we can find again and again an ordering of the leaves on a stem which marks out, on its ascending spiral, the points of a pentagram, in such a way that the sixth is above the

6. VENUS

first. This is the pattern, as we have seen, traced out by successive superior or inferior conjunctions of Venus. And in the rose and nearly all members of the rose family, this form is carried into the blossom, where we see the five-pointed star most beautifully expressed.

When poets and mystics speak about the rose, they approach in their various ways the mysteries of Venus and of the Sun. What the rose says in its form, its colour and its fragrance, the love of man for God and for his neighbour can come to resemble. But there are many, many trials for man, before he frees himself from illusion about his capacity for love. What seems to be love, is very often self-love in disguise. On the pilgrimage towards the true spirit of Venus it may be necessary to say again and again, 'What I believed I had found, is lost to me again, and I must look afresh.' Not for nothing is it described in the Gospels that from the personality who bears in particular the mystery of the true Venus, Mary Magdalene, seven devils — seven mistaken loves — had to be driven.

Everything that is beautiful, everything through which love begins, has this double character; it can lead us on or hold us back. It is deeply rooted in human nature to seek to possess, to hold still, what is of its nature changing. One of the first conflicts to be recorded in detail, the siege of Troy, has for its theme the possession of Helen. And even within genuinely loving relationships, we find again and again through possess-iveness the element of conflict. Venus wears the helmet of Mars.

In the fifteenth and sixteenth centuries there were a number of artists, poets and philosophical writers who thought deeply about such matters, and expressed them, more or less enigmatically, in their work. (Some

51

illuminating examples are shown by Edgar Wind.) Sometimes their treatment led to all-too-learned obscurity. But they could also be treated with a delicate apparent simplicity, so that their presence is seldom realized. Shakespeare in his early work refers often quite explicitly to the Martian quality in Venus — but he also exemplifies it wonderfully in such plays as *Much Ado About Nothing* and *The Taming of the Shrew*, where real loves are for a long time disguised as conflicts — or perhaps won from conflict.

But later Shakespeare is more concerned with the way in which Venus comes to exercise the healing powers of Mercury. The transition from one to the other underlies the whole of a play which is often felt to be unsatisfactory, because this is so little understood: *All's Well that Ends Well*. Here a 'Helena' comes not as the instigator of conflict but as healer and pilgrim. And we find the same uniting of qualities appearing in later plays, for example, in the figures of Marina in *Pericles* and Imogen in *Cymbeline*. In this, as in so much, Shakespeare shows his prophetic power; for all loves need to find more and more, in our time and into the future, their meaning as powers of healing.

In our time the festival of Michael begins while the Sun is in the constellation of the Virgin — but it is in the *sign* of Libra, the Balance, one of the homes of Venus, according to the long-established tradition. The Archangel Michael has often been represented as bearing the earnest scales of judgment. Our powers of love are indeed to be weighed in the balance; and it will be seen what can endure, as of abiding value to the world of which we are a part.

7. The Moon

In the fourth century BC, Aristotle felt it necessary to justify his intense interest in the behaviour of animals. He did this in a remarkable way:

> Doubtless the contemplation of the heavenly bodies fills us with more delight than we get from the contemplation of lowly things; for the Sun and stars are born not, neither do they decay, but are eternal and divine. But the heavens are high and afar off, and of celestial things the knowledge that our senses give us is scanty and dim. On the other hand, the living creatures are nigh at hand, and of each and all of them we may gain ample and certain knowledge if we so desire. . . Then will nature's purpose and her deep-seated laws be everywhere revealed, all tending in her multitudinous work to one form or another of the Beautiful. *(De Partibus Animalium.)*

Thus Aristotle was sure that when we contemplate the Sun and stars we come nearer to the divine than when we study animals. And he seems to indicate that there are other ways of learning about the stars than through the scanty information given by the senses. But even in his time, and in the following centuries, it was not easy to relate the movements of the planets, for example, to the gods whose names they bore. The course of the planets through the stars, with their alternating withdrawals, loops and advances, seemed painfully like those of drunken men; and for a long time it seemed to be the task of the astronomers to explain these in terms of the perfect circles appropriate for the divine.

Today the information achieved through the unaided senses has been immensely amplified. And now it seems for the astronomer that the stars and planets are in no sense eternal or divine. Each star is seen as having its 'proper motion', so that the shapes of the constellations are very slowly changing. Whole galaxies move in relation to each other, and themselves rotate. Hypotheses are constructed about the origin and history of every star and of the solar system. And perhaps most daunting of all, for the conception of the heavenly bodies as something divine, has been what we have been shown (and even walked upon, in the case of the Moon) as lunar and planetary surfaces.

As the event of the first human steps upon the Moon's surface withdraws a little into the past, it is still difficult to see it in perspective. Some years before it happened in 1969, the blind French writer, Jacques Lusseyran, who died in 1971, gave the warning that by stepping on the Moon we would only add to the number of countries that we see with our physical eyes but do not understand. The grim surfaces of the Moon, and of the planets that have been photographed from close at hand, have probably added to the burden of materialism in the human soul. They have made the presence of life and of man upon the Earth seem all the more freakish and accidental in an inhospitable universe. They have widened the gulf between the astronomer and the physicist on the one hand, and the poet on the other. The years since the Moon landing seem to have supported the view that spacecraft can be useful for communications and military purposes, but for very little else. At the time of the Moon flights many

7. THE MOON

people said that the enormous expenditure of resources and human ingenuity which they absorbed could have been much better used. Since that time men's minds have if anything grown more confused about the right employment of people and things. But every earthly thing, however sombre and lifeless it may appear, and every heavenly body, have their spiritual auras which the clairvoyants and initiates of every age have been able to see. And we may struggle through to the recognition that the movement of every body, whether on Earth or in the sky, manifests a soul — though they may be much more difficult to interpret than human movements. We need only to rid ourselves of the belief that we already grasp them completely if we think of them in terms of gravitation, inertia, and the like.

Even the externally visible movements of the Moon and other heavenly bodies, though they are calculable, are much less regular and machine-like than is widely supposed. From schooldays onwards, most people are satisfied, if they think about it at all, by picturing the Moon as simply going round the Earth, quite forgetting that the Earth itself is moving within the whole solar system, that the solar system is moving within the galaxy, that the Moon is sometimes nearer and sometimes farther from the Earth, sometimes more rapid and sometimes slower — not at all like the end of a spoke on a rigid wheel. Looking at the Moon without any fixed physical notions in one's mind, one could well have the same joy as in contemplating a great and graceful dance. Even what can be seen quite easily on a few clear evenings is rich in beauty and variety; and this experience of beauty is not just a subjective human emotion, but belongs as much to the whole reality as the quantitative facts.

Let us think of the sequence of events through the phases of the Moon in very simple outline. First we may picture the fine sickle of the Moon as we may see it quite soon after sunset in the western sky. It may be compared to an uptilted boat, or to a bow; as a bow, it points in the direction of the Sun, which has now gone beneath the horizon. Very soon the Moon too sets in the west. If we see it on the next evening, the illuminated part of the Moon will have become a little thicker, and its height above the horizon will be greater at the corresponding time after sunset. It will thus set later. And on each successive evening more of the Moon is illuminated, and its angular distance from the Sun is greater. We come to the moment, First Quarter, when in the northern hemisphere the entire right-hand half of the Moon is bright, like a D, and it stands at sunset no longer to the west but to the south. It now dominates the first part of the night.

During the next week, we approach Full Moon. Each evening the Moon becomes visible further east; more than half of the surface is illuminated, and more of the night. The angle between Sun and Moon, as viewed from the Earth, is now more than a right angle; each twenty-four hours this increase covers about one-seventh of a right angle, about 12°, or twenty-four Moon-breadths. Thus Full Moon is reached, when Sun and Moon are opposite one another in the sky. As the Sun sets, the Full Moon is rising in the east. It shines now through the whole night, setting about dawn.

And now the decrease of the Moon begins. There is a moonless interval after sunset; and when the Moon rises, an increasing part of the leading (upper or right-hand) side is darkened. A week after Full Moon it has reached the Last Quarter, rising only in the middle

7. THE MOON

period of the night, and standing to the south at dawn. Only the left-hand half is illuminated. During the following nights it appears nearer and nearer to the sunrise, becoming once more a fine bow, directed at the approaching Sun. Finally it becomes invisible altogether in the light of dawn, to appear again about three days later after sunset once more. The whole process has lasted just over four weeks — twenty-nine-and-a-half days.

It may seem unnecessary to describe this familiar sequence of events in such plodding detail. But tests among small groups of town-dwellers often show how shaky our grasp of these things has become. All too easily we content ourselves with the thought that the Moon goes round the Earth, and leave it at that. But through this our thinking about the Moon hardly comes into activity at all. And our feeling may only begin to stir when we realise what a great difference of quality there is in our impression of the Moon when it is close to sunset or sunrise, and can become the companion of the evening and morning stars, Venus and Mercury, and the Moon when it can rule the mid-period of the night, and be with Mars or Jupiter or Saturn during their times of greatest brilliance.

Our picture is very much enriched when we consider the Moon's movement not only in relation to Sun and Earth, but also in relation to the stars. Like that of the Sun, the path of the Moon lies within the constellations of the Zodiac, though the paths are not the same, but slightly at an angle. (If they were not, we should have eclipses of the Sun and Moon every month, instead of at certain periods of the year.)

As the waxing Moon increases its distance from the Sun up to First Quarter, it passes through the constellations which the Sun will reach during the next three months. At Full Moon, plainly, it is in the constellation opposite to that of the Sun. This has a significant consequence. When in winter the Sun follows a low path, for example in Capricorn in the later part of January, the Full Moon follows a high course like that of the midsummer Sun. In Britain the Full Moon at the beginning of the constellation of the Crab is above the horizon for nearly fifteen-and-a-half hours. Correspondingly the Full Moon around midsummer follows a low course and may be above the horizon eight hours or less. In spring and autumn the Full Moon is in a middle position, and makes an arch similar to that of the Sun at the same time. All this too need not remain a bare geometrical conception; it can become part of our feeling, that the Moon in midwinter brings a reminder of summer, and correspondingly through all the seasons. The young Moon is then like a prophet of the season that is coming, and the old Moon as it wanes, a rememberer of the season that is past.

The Moon always turns almost the same face to the Earth; it rotates once on its own axis while moving round the Earth, like a respectful lady in waiting going round a queen. Owing to oscillating movements called the 'libration' more than half of the Moon's surface is visible from the earth, but about two-fifths had never been seen by human eyes until recent space flights. Correspondingly, a dweller on this far side would never see the Earth; for fourteen Earth days he would be able to gaze out into a starry heaven of absolutely undisturbed clarity. For the rest of the month the Sun would be moving among the stars, but without the effects produced by our atmosphere. And a dweller in the middle of the near side of the Moon would always have the Earth high in the heavens above him, never disappearing below his horizon, tremendously bright among the stars at 'Full Earth', dark near the Sun at 'New Earth' (and sometimes eclipsing the Sun, when on Earth there is an eclipse of the Moon).

This whole description has been attempted from the aspect of human physical sight and human conceptions of space. It leads us perhaps towards some enrichment of our feelings about the complex relationships of Earth, Moon and stars. Ever looking on one side to the Earth, ever looking on the other towards a wide universe in which no Earth is visible, receiving and giving away the bounty of the Sun's life; the Moon may awaken in us the picture of something like a great cosmic door between the wide heavens and the life of Earth. What echoes to us from the ancient world, and what we can learn from the initiate research of Rudolf Steiner may confirm such a picture.

Rudolf Steiner speaks not only of the Moon as a heavenly body, but also of an entire invisible sphere round the Earth, of which the visible Moon is something like a frontier marking-stone. It is in this sphere that human souls prepare themselves for the two great transitions — from spiritual existence into earthly life, and from earthly life into spiritual existence.

We have been citizens of the spiritual world, beyond space and time as we experience these on Earth. The beings of all the hierarchies were our companions, and we lovingly shared in their creative life. We sounded in the heavenly Word and shone in the Light of God. And from the spiritual heights we turned towards the far Earth. There a task waited for us; to take it up, we had to pass through a great transformation. And at last we found in the Moon sphere the powers that would help us to plunge into the stream of earthly life, and enter into the form which earthly parents could give. And at this time we could meet those who had not long been parted from their earthly body, whose souls show strongly the effects of their personal destiny.

Each of us during our lives, even if we feel a great longing for spiritual existence, becomes to some extent an earthbound creature — attached and accustomed in all sorts of ways to limited surroundings and a comfortable routine. And under the influence of these things we have left undone much that we came on to Earth to do. The Moon sphere has to receive us as we are, and prepare us for the other great transformation, that we may learn to seek again our eternal being. The facts of the past life, as they appeared not to our eyes but in the sight of spiritual beings, have to be recognized with all their consequences. Sharing the patient contemplation of wise beings, we find new powers of will awakened in us.

When souls on their way towards earthly life see how much purification is needed by the souls who have left it, does this give them the impression that the Earth is a place on which nothing is to be found but failure and guilt? We can think of the astronauts when they know themselves to be heading back towards Earth with its breathable atmosphere, its rivers and trees and living creatures. Is the experience of souls utterly different?

When Saul came down from the desert uplands into the wide fertile gardens around Damascus and the grey gleaming olive trees, he bore in his soul an austere and indeed tragic Moon-religion. And he met the grace and mercy of the living Christ. He found all that gave the Earth meaning and hope. It is part of the task of a living Christianity to develop on the Earth a spiritual

7. THE MOON

light, which will shine in those who leave it, and for those who approach it. The better the deed of Christ is understood and served on Earth, the more we shall see the work that is done in the sphere of the Moon not as that of stern taskmasters, who bind men to their destiny, but a blessing and grace for souls on either path.

The Zodiac

The *signs* of the zodiac are twelve equally sized segments beginning at the spring equinox (0°). The *constellations* are made up of the visible stars, and vary in size. Due to the slow movement of the spring equinox against the stars they begin in a different point from the signs. The beginning points are given for 1985.

Latin name	English name	sign begins at	constellation begins at
Aries	Ram	0°	29°
Taurus	Bull	30°	53°
Gemini	Twins	60°	89°
Cancer	Crab	90°	117°
Leo	Lion	120°	138°
Virgo	Virgin	150°	173°
Libra	Scales or Balance	180°	219°
Scorpio	Scorpion	210°	237°
Sagittarius	Archer	240°	268°
Capricorn	Goat	270°	298°
Aquarius	Water Bearer	300°	326°
Pisces	Fishes	330°	351°

Bibliography

Aristotle, *De Partibus Animalium.* (Trans. D'Arcy Thomson).

Bühler, Walter, article in *Goethe in unserer Zeit,* Basel, 1949.

Rihouët-Coroze, S, *Biographie de Rudolf Steiner,* Triades, Paris, 1951.

Frieling, Rudolf, *Hidden Treasure in the Psalms,* Christian Community, London, 1967.

Lusseyran, Jacques, *And There Was Light,* Floris, Edinburgh, 1985.

Schultz, Joachim, *Rhythmen der Sterne,* Philosophisch-Anthroposophisch, Dornach, 1977. To be published in English, Floris, Edinburgh 1985/86).

——Article in *Goethe in unserer Zeit,* Basel, 1949.

Sitwell, Edith, *Collected Poems,* Macmillan, London, 1957.

Vreede, Elisabeth, *Astronomie und Anthroposophie,* Philosophisch-Anthroposophisch, Dornach, 1980.

Wachsmuth, Günther, article in *Goethe in unserer Zeit,* Basel, 1949.

Wind, Edgar, *Pagan Mysteries of the Renaissance,* Oxford University Press, 1980.